UNDER THE SEA

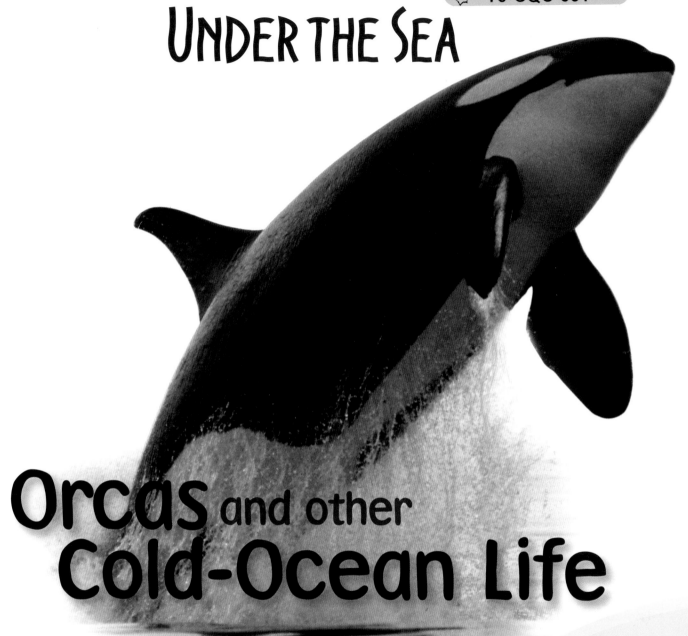

Orcas and other Cold-Ocean Life

Sally Morgan

QEB Publishing

First published in the United States in 2008 by
QEB Publishing Inc.
23062 La Cadena Drive
Laguna Hills, CA 92653

www.qeb-publishing.com

Library of Congress Number: 2008010037

ISBN 978 1 59566 680 2

Author Sally Morgan
Consultant Camilla de la Bedoyere
Editor Sarah Eason
Designed by Calcium
Picture Researcher Maria Joannou

Publisher Steve Evans
Creative Director Zeta Davies

Printed and bound in China

Picture credits
Key: T = top, B = bottom, C = center, L = left,
R = right, FC = front cover, BC = back cover

Corbis Amos Nachoum 13B, Brandon Cole FC, 1, 6-7,
Paul A Souders 10B, Stuart Westmorland 21B,
Theo Allofs 7B
Dreamstime 18L, 18R, 19BL, 19T
Getty Images Aurora/Sean Davey 14B
istockphoto 7T, 8, 9T, 12L
Photolibrary Duncan Murrell 10-11, Herb Segars 19BR,
James Watt 8-9, 10T, 20-21, Mark Stouffer 15R,
Harold Taylor 5, Oxford Scientific/Richard Herrmann 3,
14-15, 16-17, 16L, 17B, Rodger Jackman 12-13
Shutterstock 2-3, 4-5, 4L, 4R, 22-23, 24

Words in **bold** can be found in the
glossary on page 22.

Contents

Cold oceans

The oceans are huge! They cover nearly two-thirds of the Earth's surface. The oceans are deep, too. In some places they reach nearly 7 miles below the surface.

Most of the Earth's surface is covered by water.

Down in the ocean's depths, the water is very cold and dark. Nearer the surface, water is usually warmer because it is heated by sunlight. Most animals live at the surface.

African penguins hunt for fish in the cold waters of the Atlantic Ocean.

Animals of every shape and size live in salty ocean waters. The massive blue whale lives alongside tiny **microscopic** creatures called **plankton**. All ocean animals are suited to living in their watery world.

Tiny plankton need to be **magnified** for us to see them.

The ocean may be full of many more amazing animals, yet to be discovered.

Orcas

The orca is also called the killer whale, even though it is actually a dolphin. It is one of the fastest ocean hunters. The orca's black-and-white markings make it easy to identify.

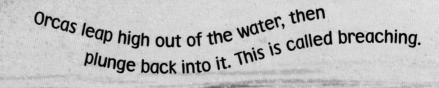

Orcas leap high out of the water, then plunge back into it. This is called breaching.

Some orcas stay in the same pod for their whole lives.

Orcas live in groups of about 30, called **pods**. Each pod has its own special language. Orcas "talk" to each other most of the time, so life in a pod can be very noisy!

Orcas will even swim onto the beach to catch a seal.

Dolphins

Dolphins are intelligent and **acrobatic**. Their slim, sleek bodies slip easily through the water. These playful animals often leap out of the water or swim alongside boats.

Dolphins can leap high out of the water. They can also spin in the air and perform somersaults!

Like killer whales, dolphins swim in groups called pods.

Dolphins use sound to find food. They squeak and whistle. These sounds travel through the water and bump into other animals in the water. This creates **echoes**. The dolphin uses the echoes to work out where its **prey** is.

Sound

The humpback whale is one of the giants of the ocean. This heavy animal is also a strong swimmer, singer, and acrobat!

Humpback whales jump from the water and perform amazing leaps and twists in the air.

Dolphins have small, sharp teeth, which they use to eat fish.

Humpback whales

The humpback whale is one of

huge,

Whales give birth underwater. They push their young to the surface so they can breathe.

The humpbacks talk to each other by singing. Each whale has its own special song, made up of clicks and whistles.

Humpbacks open their mouth wide to take in lots of water. They eat tiny animals and plants, called plankton, that live in the water.

Whale sharks

The whale shark is the largest fish in the ocean. Some grow as long as a bus and weigh a massive 16.5 tons.

The whale shark feeds by swimming along with its mouth wide open. It takes in huge mouthfuls of water, which it sieves through its **gills**. All the fish, squid, krill, and plankton in the water are eaten.

The whale shark's mouth is 4.5 feet wide, large enough to swallow a child!

Female whale sharks do not lay eggs. Instead, they give birth to as many as 300 baby sharks at one time.

The whale shark has a beautiful pattern of spots on its body.

Herring

The herring is an important fish in the **food chain**. It eats tiny animals and plants called plankton. In turn, larger fish, such as tuna, eat the herring.

Herring swim together in large groups.

Thousands of small, silvery herring form huge **shoals**. In the day, they hide from **predators** in the dark, deep water. At night, they swim to the surface to feed in the safety of darkness.

Herring are covered in shiny scales that glint in the sunlight.

Each female herring lays about 40,000 eggs. Only a small number of young herring live to be adults—most are eaten by predators.

Orcas hunt herring shoals.

Tuna

The tuna's long, sleek body is shaped for speed. As its tail **fin** cuts through the water, the tuna folds back its other fins to create an even smoother shape. This helps it to reach speeds of up to 43 miles an hour. That's as fast as a racehorse can run!

Tuna hunt and eat smaller fish.

Tuna live and hunt in large shoals. They swim long distances in search of food. They often travel more than 6200 miles across the oceans.

The tuna's sleek shape helps it to speed through the water.

Up to 1000 tuna fish may swim in one shoal.

Sunfish

The sunfish looks like a fish with half a body! Instead of a tail fin, its rounded body flattens at one end. Large fins at the top and bottom of its body help it to steer and stay upright.

The sunfish's large, round mouth is perfect for swallowing jellyfish. Unfortunately, many sunfish die after eating plastic bags, which look like jellyfish floating in the water.

The sunfish's mouth is always open and ready to catch its prey.

The sunfish is one of the largest fish in the ocean.

This massive fish can grow 13.8 feet tall—taller than two men. It weighs up to 4409 pounds—about as heavy as 25 men!

Sunfish sometimes swim in groups of up to ten fish.

Jellyfish

The jellyfish is not a fish. It is an **invertebrate**—an animal without a backbone. The water supports the jellyfish's floppy body so it floats, carried around by the ocean's **currents**. If a jellyfish washes up onto a beach, it collapses into a soft blob.

A jellyfish's long tentacles trail behind its body.

A group of jellyfish is called a swarm.

Jellyfish live near the surface of the ocean.

The jellyfish's long, hanging **tentacles** are covered in tiny stings that explode with **poison** when touched. Some jellyfish stings are deadly to people.

Glossary

acrobatic able to twist, turn, and jump easily

current the flow of water that moves through the ocean

echo the noise made when sound bounces off objects

fin the part of a fish used to swim and steer

food chain plants and animals that feed on each other. For instance, plants are eaten by fish, fish are eaten by seals, and seals are eaten by orcas

gill an opening in its body through which an underwater animal breathes

invertebrate an animal without a backbone

magnify make something look much bigger than it really is

microscopic too tiny to see without an instrument called a microscope

plankton small plants and animals that float in the surface waters of the oceans

pod a group of dolphins or orcas

poison something that can harm or even kill

predator an animal that hunts other animals

prey an animal that is hunted by other animals

shoal a group of fish

tentacle a long, armlike limb of a sea creature. It is used for feeding and holding, or even stinging

Index

Ideas for teachers and parents

- Visit an aquarium to see fish up close and learn about their life cycles. At some larger visitor attractions you can see orcas and dolphins.

- Visit a fishmonger's or a fish market to see the different types of fish on sale.

- Visit a fishing harbor to watch the fishing boats unloading their catch. You can find out more about the different types of net on the Internet.

- Find out how dolphins are trained to help the navy with tasks including alerting the navy to the presence of underwater mines.

- Make a drawing of an ocean food web. Use information in the book and from the Internet to find out what eats what. For example, plankton are eaten by herring, herring are eaten by tuna and orca. On a large sheet of paper, draw outlines of the animals. Link them together with arrows to show the feeding relationships. Then color in the animals.

- The ocean is under threat from overfishing, global warming, and many forms of pollution, such as trash, sewage, and oil spills. Find out more about these threats from books and the Internet.

- Encourage children to think up fun stories and poems about the orca and its life in the ocean.

- Make a word search using the different ocean-related vocabulary in this book.